HYDROPONICS

101

The Easy Beginner's Guide to Hydroponic Gardening. Learn How To Build a Backyard Hydroponics System for Homegrown Organic Fruit, Herbs and Vegetables

Tommy Rosenthal

© **Copyright 2020** by Tommy Rosenthal – All rights reserved.

In no way is it legal to reproduce, duplicate, or transmit any part of this document in either electronic means or in printed format. Recording of this publication is strictly prohibited and any storage of this document is not allowed unless with written permission from the publisher.

The information provided herein is stated to be truthful and consistent, in that any liability, in terms of inattention or otherwise, by any usage or abuse of any policies, processes, or directions contained within is the solitary and utter responsibility of the recipient reader. Under no circumstances will any legal responsibility or blame be held against the author for any reparation, damages, or monetary loss due to the information herein, either directly or indirectly.

The information herein is offered for informational purposes solely, and is universal as so. The presentation of the information is without contract or any type of guarantee assurance.

REVIEWS

Reviews and feedback help improve this book and the author.

If you enjoy this book, we would greatly appreciate it if you were able to take a few moments to share your opinion and post a review.

Table of Contents

Introduction 6
Chapter 1: What is Hydroponics? 10
Chapter 2: How To Choose The Right Hydroponics System 14
Chapter 3: Best Fruits and Vegetables For Hydroponic Gardening 26
Chapter 4: Things To Consider For Hydroponic Gardening 34
Chapter 5: How To Design Your Own Hydroponics System 44
Chapter 6: How To Take Care Of Your Hydroponic System 58
Chapter 7: Tips and Tricks For Success 68
Conclusion 80
BONUS CHAPTER: What is Aquaponics? 84
Did You Like This Book? 96
By The Same Author 97
Notes 100

Introduction

Key Takeaway: *Hydroponics is the art of growing plants without soil. Not to be confused with aquaponics, this type of growing can yield faster harvests and puts all of the control in the hands of the farmer.*

Thank you for picking up this copy of *'**Hydroponics 101**: The Easy Beginner's Guide to Hydroponic Gardening. Learn How To Build a Backyard Hydroponics System for Homegrown Organic Fruit, Herbs and Vegetables.'*

If you are considering this book, then it is because you have found an interest in hydroponics and wish to know more about it.

Hydroponics is the art of growing plants without soil. Do not confuse this with aquaponics, which is the science of farmed fishing and utilizing the fish's waste product as fuel for their hydroponics. Growing plants without soil is a rising trend among many in today's culture: it removes the dependency on soil and does away with the confusion of crop rotations in order to preserve soil health and allowing it to rejuvenate after a harvest. With hydroponics, the most common stereotype is the envisioning of plants suspended in air while their roots sit in a vat of water. While this is a form of hydroponics, called the nutrient film technique, it is not the only one available to those researching their options.

Whether you are looking to educate yourself further on the material or you are looking for a broad checklist to make sure you have everything you need to get going, this book will provide it for you. It has basic explanations of everything you need to know about the subject, plus a simple project breakdown of how to build your very first hydroponic system. Keep in mind, a project like hydroponics has a learning curve, and it is best to start small and work with a few plants at a time to perfect your craft than it is to go all-in with a sprawling system and no prior knowledge of how it works.

We will cover everything from the types of hydroponic systems available all the way down to the easiest plants to start growing hydroponically. It will cover materials necessary in order to make your own hydroponic system, and it will talk you through several basic how-to's when it comes to taking care of your plants and maintaining your basic system. By the time you are done with this book not only will you have a thorough understanding of the systems available to you, but you will have all of the information you need to get started building your own system so you can begin growing your own fruits and vegetables.

I hope you are excited!

Let's get started, shall we?

Chapter 1: What is Hydroponics?

Key Takeaway: *Hydroponics is the art of growing plants without soil.*

So, you want to give hydroponic growing of plants a try. Wonderful! But what exactly is hydroponics?

Hydroponics Explained

Hydroponics is the art of growing plants without soil. The word "Hydroponic" actually means "working water" in Latin. There are many methods of hydroponic gardening. The plants can be grown in gravel, liquid, or sand. Nutrients can be added, but the key thing to remember is that the plants are grown without soil.

Why Grow Plants Without Soil?

The first thing you need to understand is why plants "require" soil in the first place.

Soil is important to the growing and nurturing of plants because the soil acts as a medium for growth. A medium for growth is simply the substance that bears the weight of the plant and keeps it upright while the root system does its job. The soil is able to store the required nutrients the plant needs, and it serves as a mode for balancing normal pH levels in the roots of the plant.

If the pH levels in the plant's roots are unstable, it begins to break down the roots which can inhibit the plant's ability to absorb the required minerals it needs to thrive. This imbalance is one of the highest causes of plant death among those attempting to flex their green thumb, and it is what causes many individuals to give up.

However, when you give a plant what it needs when it needs, in the amount it requires, the plant can thrive no matter what its roots are submerged in.

In hydroponics, that medium for growth is water, and the nutrients come from carefully measured and organically-produced products that are combined with the plant's water source.

With hydroponics, the roots are given a break. In soil, the roots have to branch out and search for the nutrients, which is how all plants obtain their soiled root systems. In water, the dissolved proportion of nutrients do not have to be sought out, and the energy that is usually expended developing the roots is now spent developing the plant itself, which can aid in the speed in which the plant grows and produces fruits and vegetables.

There are six basic systems in hydroponics. We will discuss those next.

Chapter 2: How To Choose The Right Hydroponics System

Key Takeaway: *There are six different systems from which all other professional systems derive themselves from, and each system has its own unique positives and negatives.*

In hydroponics, there are six basic systems:

- drip system
- nutrient film technique
- water culture
- wick
- ebb and flow, and
- aeroponic

Within these six systems, many individuals have come up with and engineered hundreds of variations, but all of their methods stem from these six fundamental systems. Each system has their ups and downs, and some systems are more popular than others because of their ease and adaptability to a home, but all work towards the same purpose: enabling plants to thrive with roots submerged in water.

Let's take a closer look at each system, so you can decide which one would work best for you!

Drip System

The drip system has two different types:

- recovery
- non-recovery

This is the most widely utilized hydroponic system around because of its simple operation and timer-controlled pump. The only difference between the recovery and non-recovery methods is how the runoff from the drip line is treated: in the recovery drip system, the runoff is collected and reused, and in the non-recovery drip system, the runoff is not.

The operation for this system is very simple: there is a timer that controls a pump that is completely submerged in water. This timer activates the pump which provides the nutrient solution stored within it. This nutrient solution travels through a drip line and is, quite literally, dripped onto the base of every plant. Then, depending on the type of drip system used, the excess runoff is either collected and reused or left alone to dissipate as necessary. The growth medium for a drip system would be something akin to river rocks because it aids in the natural draining of the water dripped onto the plants and provides a sturdy foundation for the plant to sit on.

With a recovery system, you need to keep in mind the pH balance of the nutrients within the original reservoir. With a non-recovery system, the excess solution does not get recycled so the pH levels within the nutrient solution keeps a stable pH level. When recovering and recycling that nutrient water that has already been used, there can be massive shifts in the pH level of the nutrient solution which, if not caught, can lead to massive root system deaths in your plants. If you choose to use the recovery system, make sure to periodically check and adjust the pH levels as necessary.

Nutrient Film Technique

The nutrient film technique (often known by its abbreviation "N.F.T.") is the system that pops into most people's minds when someone says the word "hydroponics." This system utilizes a constant flow of nutrient solution, so the timer for a pump is abandoned in this setup. Instead, this mixed solution is filtered into a growing tray of some sort and left to flow organically over the submerged plants. Then, all of this drains back into a reservoir for collecting. In this system, no growing medium is utilized, which helps with expenses because there is nothing to change out after the harvesting of every crop. In this system, the plant is supported by a small basket of some sort and has its roots dangling in the nutrient solution that is flowing freely.

However, keep in mind that these types of systems are incredibly prone to pump failures and power outages. Many people also experience rapid root dry-out when something impedes the natural flow of the water, so the system has to be checked multiple times a day in order to ensure all things are running smoothly. Many people argue that this is the system that requires the most daily maintenance.

Water Culture

Water culture is another system that is utilized, and it is the simplest of all the hydroponic systems. There is a basic platform that holds the plant, usually made from styrofoam, and it floats right on top of the nutrient solution. Then, a pump supplies air to the air stone, and this air stone then bubbles the mixed nutrient solution towards the roots of the plant. Not only that, but this bubbling also supplies necessary oxygen to the root systems as well, aiding in root health even though they are almost fully submerged.

The only downside to this type of system is that unless the plant is a water-loving plant (such as lettuce), the plant will not do well in a constantly-submerged state. But, the upside is that the constantly-submerged state eliminates the need for any growing medium, which makes this system even more cost-effective than most.

This is the type of hydroponic system that is most popular with teachers in their classrooms. It's incredibly cheap to create and requires minimal maintenance other than checking the pH levels of the water and nutrient mix. Keep in mind though that larger or more long-term plants will not work well with this type of system.

Wick System

The wick system is considered the simplest system because of its passive nature. Essentially, this system has no moving parts. What happens is the nutrient solution that is mixed by you is pulled into the growing medium from some sort of reservoir with a wick. The wick works by simply absorbing the nutrient water and traveling it up the fibers of the wick into the small reservoir the plants are sitting in. If you have ever watched a piece of string quickly soak up water it is dipped into, that string is "wicking" the moisture from one end to the other. The wick system of hydroponics utilizes this passive movement to introduce the nutrient solution to the plants. Some of the more popular growing mediums for this type of system are coconut fibers and vermiculite, because they are the most susceptible to the wicking process this system requires.

For this system, smaller plants (such as herbs) and plants that do not need as much watering will be ideal.

Ebb and Flow System

The ebb and flow system (sometimes called the "flood and drain" system) works under the premise of flooding a growing tray with a nutrient solution before quickly draining it back down into the reservoir. This is another active system, and the flooding and draining are controlled by a submerged pump connected to a timer. When the timer turns the pump on, the solution flows into the growing tray, and when the timer shuts the pump off, the solution drains back into the holding tank. This timer has to be set to activate and deactivate multiple times a day, and those "multiple times" are gauged by the types of plants being grown and how big their root systems are.

This system is considered to be the most versatile system simply because a variety of growing mediums can be utilized, however, it is susceptible to power outages and pump failures that can increase the risk of the roots drying out. One way to combat this is to utilize a growing medium that retains more water than most, such as rockwool or coconut fiber.

Aeroponic

The last system for hydroponics is called aeroponic. Out of all the systems, passive and active, this is the most technical system available. Just like the N.F.T. system, the majority of the growing medium is air. But, the nutrients are misted onto the plants. These mist blasts are done every few minutes because, just like the N.F.T. system, the roots are fully exposed and can dry out easily if not consistently sprayed down with water. The most massive drawback to this is, yet again, roots drying out: if the timing of the mists is interrupted for any reason, the risk of plant life dying off is high.

A timer is set up to control the nutrient pump like most of the other systems, but the difference is that this system needs a much shorter timer cycle. Some need to be set as often as every few seconds, but all of the timer settings never breach more than a few minutes between each mist. The plants are suspended with the roots exposed either in little baskets or in growing cubes, and you have to make sure these plant holders do not become saturated. Too much saturation that gets held against the stem of the plant can lead to stem rot and plant death.

How to Choose The Right Hydroponics System For You

All of these types of systems have their own strong suits as well as their drawbacks. With any system that requires the plant to be suspended within a holder, you have to make sure that holder is not becoming saturated with water and being held against the stem of the plant for too long. Stem rot is one of the largest triggers of plant death in the world of hydroponics, and it is imperative to make sure the nutrient water, no matter where it is coming from, is draining off in a timely manner.

This is why many people choose the growing mediums they do: the material used in place of the soil either helps the draining of the water or the holding of the water, depending on what system is being used. If you know the purpose for your growing medium, it eliminates potential items to consider in order to make the choice between growing mediums easier.

With all of these systems at your disposal, picking and choosing one requires asking yourself some questions: what is your skill level? What type of maintenance work can you handle? How big do you want it to be? Where will the garden be planted? What plants do you want to grow? How big do those plants get when they are fully grown? All of these questions will carry a substantial amount of weight when choosing the right system to use for your garden, so it is best to answer these questions now before choosing one to dive right into.

However, once you choose a system to use for your plants, the worst of it is out of the way. What remains is educating yourself on the system you have chosen, the plants you want to grow, and the nutrient solution for the water you will use to sustain the plant life you want.

Each of these systems have their pros and cons. With those upsides and downsides come plants that grow in hydroponic atmospheres better than others. The good news is this: those that are better suited for hydroponics come with fun little tips and tricks to help with the feeding schedule and the mixing of the nutrient water for these specific plants.

What does all of this mean? It means a successful first harvest, even if you do make glaring mistakes.

Chapter 3: Best Fruits and Vegetables For Hydroponic Gardening

Key Takeaway: *There are eight fruits and vegetables that grow better in hydroponic gardens than any other ones. These plants will help a first-time grower to reap their first harvest, implanting confidence they can use to try again without their mistakes holding them back.*

Even though hydroponics can be utilized to grow a vast range of fruits and vegetables indoors without soil, there are some plants that thrive better under these conditions than others. Many plants that are designated for beginners are labeled this way because of their rapid growing abilities because of the dense amount of nutrients available to them. This means that, because they grow rapidly, it takes significantly less time to spot a problem and fix it rather than living out the life of the crop before realizing it will not bloom. Then, if the plant does die off, it takes little to no time at all to sprout another plant and begin again.

The small window of trial and error with many fruits and vegetables is ideal for beginners, especially when tweaking growing and feeding methods. What usually takes farmers years can take someone months, or even simply weeks!

There are eight fruits and vegetables that are easiest to grow hydroponically:

1. Lettuce
2. Spinach
3. Cherry Tomatoes
4. Cucumber
5. Peppers
6. Strawberries
7. Mint

8. Basil

They are wonderful beginner plants to work on when developing your own technique and choosing the system you ultimately wish to use.

Let's take a closer look at each one!

Lettuce

The first plant is lettuce. This is, by far, the most popular choice among beginning gardeners utilizing the hydroponic technique. The reason lettuce is the easiest and most popular choice is because it is a water-heavy plant. It requires a great deal of water in order to keep the plant sustained and hydrated, so "stem rot" and drowning the root system is very hard to do. This massive requirement of water also means it requires very little upkeep once it begins to grow. Many people enjoy being able to harvest the outer leaves instead of the entire head of lettuce, which gives people a continuous supply of lettuce without having to plant multiple baskets of them.

Spinach

Spinach is another plant that is easy to grow hydroponically. It is yet another plant you can harvest leaf by leaf, and it is another plant that requires a lot of water in order to keep it growing and flourishing. Like lettuce, spinach also needs very little upkeep once the root system begins to grow. Other leafy greens that fall into this category are watercress (which is a plant that thrives on constantly moving water), kale, and arugula.

However, with anything that is leafy and green, make sure it does not become too big to handle. Lack of air circulation between the leaves of a plant like the ones mentioned above can lead to suffocation and leaf death, which can aid in overall plant death if left alone for too long.

Cherry Tomatoes

The easiest plant that is not a leafy green are cherry tomato plants. Because their blossom-end stems need a constant supply of water in order to not dry out, they are less susceptible to the beginner's issue of overwatering plants. Just make sure to keep in mind that any sort of tomato, including cherry tomatoes, require a lot of light to grow.

Cucumbers

Cucumbers are another plant that do well with first-timers. Like tomatoes, cucumbers need enough sunlight to boost plant growth, but are easy to grow because they grow so quickly. It is very hard for things like overwatering and stem rot to affect a plant like a cucumber that grows so quickly. Keep in mind though that regular cucumbers sprout vines that will need to be propped up somehow. If you do not want to mess with the vines, opt for a bush cucumber seed to get you started.

Peppers

Peppers are yet another vegetable that grow well and easily hydroponically, but this is also another vegetable that needs its sunlight. If you are hydroponically growing in a small space, opt for a smaller pepper, such as a chili pepper, to grow your first time around. Their rapid growth rate enables even the most basic of first-timers to get at least one round out of the growth before noticing anything that might go wrong. This type of growth rate is a good confidence booster for those simply starting out, because at least one product yield is guaranteed before things like too much water and an imbalanced nutrient water become apparent.

Strawberries

There is a fruit, however, that is easy to grow hydroponically, and that fruit is a strawberry. Year-round cultivation of fresh strawberries is made possible by hydroponically growing them, and because strawberries are yet another water-heavy plant food source, the rate at which they grow is usually sped up with the direct access to nutrient-dense water. The root systems of strawberries are small, so drying them out is easier, which makes this plant less prone to the damage of over-watering as well.

Mint and Basil

But, there are also two distinct herbs that are great with beginning hydroponic gardeners as well: mint and basil. Water mint is a semi-aquatic plant that grows best in shallow waters to begin with, which makes this plant the perfect beginning plant. Peppermint is another mint leaf that grows well this way, and because they are herbs they do not require as much nutrient water as many other plants. Basil is the same way, and the great thing about growing herbs hydroponically is the fact that you can easily build your own system, active or passive, on a smaller scale to test everything out. Herbs do not need as much water as many other growing plants mentioned above, and this means that things such as system failures and irregular waterings will not affect these plants like other ones such as lettuce and escarole.

Chapter 4: Things To Consider For Hydroponic Gardening

Key Takeaway: *There are a number of things you need to consider if you want to create your own hydroponic system. Make sure you pick the right system and plants for you. Other things to consider are portability, type of water, nutrients, and pH levels.*

The reason why these plants grow so well hydroponically is because of their drastic difference in water needs. The two biggest mistakes beginning hydroponic gardeners make is either overwatering or underwatering their plants. Mixing nutrient solutions and watering plants is an art form, and it is one that is acquired over time. However, lacking that first big harvest can be very discouraging for many individuals, and it is why so many walk away from this form of farming.

Mixing a garden of herbs that require very little water and plants like lettuce and strawberries that require a lot of water will ensure a decent harvest no matter the mistake you make. Then, you can take that confidence and split the garden in half, placing the herbs in one small garden and the water-heavy plants in another and try your hand at maintaining two small gardens.

This trick is a wonderful way to boost your self-esteem for the craft and help grow your ability to stick to maintenance schedules.

Choosing The Best Hydroponic System and Plants For You

Choosing the plants to start out with is only part of the challenge. You also need to define which hydroponic system will be the best for what you are choosing to farm because some plants thrive in some systems better than they do in others. Lettuce and other leafy greens work best in a nutrient film technique system, while tomatoes and cucumbers flourish phenomenally in both recovery and non-recovery drip systems. Radishes and other beet-like vegetables grow well in a water culture system, and vegetables such as celery flourish well in an ebb and flow system.

Then, you have fruits which all grow well in their different systems, too. Melons, for example, all grow very well alongside celery in an ebb and flow system because of their hearty nutrient requirements and constant need for water, while our strawberries (well, any berry to be exact) that are so easy to grow flourish substantially in the nutrient film technique system.

But, the herbs mentioned in the last chapter (as well as oregano, sage, and rosemary) do well with either the drip system or the passive wicking system.

Do You Want a Portable Hydroponic System?

Yet another thing to take into consideration is how portable you want your hydroponic system to be. The ebb and flow system works for housing large amounts of different plants, but are completely immovable once set up. But the drip system, wicking system, or water culture system have various amounts of mobility, which is important should you be growing something that requires sunlight. If you are using an immovable system, make sure to have the proper materials necessary to induce shade when it is necessary and promote sunlight when needed.

Is The Water You Use Hard or Soft?

Mixing nutrients is not as hard as it seems, either, but there are things to keep in mind when preparing the water nutrient mixture. First off, you have to determine whether the water you use is hard or soft. Why? Because soft water is lower in its mineral count, meaning there are less variables to maintain when attempting to stabilize the pH balance. Plus, hard water usually contains high amounts of calcium, which can cause brown discharge to build up on the roots, and this can threaten plant life as a whole. But, do not make the mistake of running your hard water through a water softener and then using it. Water run through softeners is way too salty to be utilized for plant growth.

If you find out you have hard water, then there are solutions you can purchase to mix with your hard water that has a base solution that aids in leveling out pH and mineral levels in naturally hard water. Before you mix your nutrients together, mix this into your water.

However, if you test your water and you come to find out that your water is laden with minerals and calcium (above 300 ppm total solids), then you need to have a reverse osmosis filter installed on your kitchen sink, at least. Not only will this provide better drinking water, but it will ensure the highest quality water for your hydroponics plants. If the total solids in your water is greater than 300, no mixes with basic solutions will help the water you wish to soak your plants' root systems in.

If you want to take it a step further, you could even collect rainwater in a barrel or bucket to use for your garden.

Nutrients

Now, when you remove plants from soil, they need nutrients. Remember, the soil is not what gives the plant life, it is merely what suspends the plant so the roots can dig for its life-sustaining nutrients. The six macros needed for any plant are:

- nitrogen
- potassium
- phosphorus
- sulfur
- magnesium, and

- calcium

But, plants, just like the human body, also need several micronutrients in order to thrive as well. Hydroponic fertilizer solutions are not what you believe fertilizer to be, it is more like your plant's one-a-day vitamin supplement: it not only houses the six major nutrients, but all of the minor ones as well. Some of these micronutrients are:

- iron
- cobalt
- zinc
- chlorine, and
- Selenium

pH Balance

But, this is not the only requirement: the pH balance of the water is important, too. The pH level, designated by a number on a scale of 0-14, tells you how acidic or alkaline (basic) your water is. If the pH is off-balance in any way, this can hinder the growth of plants as well as cause root system death, with eventually kills off the plant. Testing it is very simple: you can purchase pH testing strips that simply dip in your water solution, or you can purchase a pH testing kit like the ones they use at swimming pools. Either is fine and just as accurate as the other.

The necessary pH range for any plant water with the proper solutions mixed in is between 5.5 and 6.5, with 6.0 being the ideal scenario and the ultimate goal to achieve. Adjusting these pH levels accordingly takes a pH upward or a pH downward solution. These leveling bottles will be labeled as such, and a little goes a long way. However, if you are in a pinch, white vinegar lowers pH balances and baking soda raises it. But, do not use those home remedies as a long-term solution because they are not buffered nor are they stable. Those are better in one-time situations, where you cannot put your hands on what you need.

No Cheat Sheet

Unfortunately, there is no true "cheat sheet" to growing hydroponic plants. It is all about a stable harness for the plant, the required growth medium, and making sure the roots are hydrated without becoming oversaturated with water. However, many plants that require heavy amounts of water or little-to-no water are good for first-timers because they can counterbalance many different issues beginners face, from over-watering to under-watering and machine malfunctions.

Once people become familiar with the concept, many wish to build their own systems from scratch. If you are handy you will need certain materials, but in the long-run building your own system can be much more reliable and easier to maintain.

Chapter 5: How To Design Your Own Hydroponics System

Key Takeaway: *Making your own system is not as intimidating as it sounds. Every system uses many of the same parts, and it is simply a matter of choosing the right system for your lifestyle and the plants you wish to grow.*

Believe it or not, it is not as intimidating as it sounds. For every hydroponic garden there are specific characteristics that are necessary in order for it to work. Just like different dentists across the nation have specific tools they all use, different hydroponic gardens are still made up of the same elements because, no matter how you plant the roots in the water, they still need a basic nutrient makeup and a way to be suspended in that water.

In order to understand how to build your own system, you need to understand all of the parts that encompass the system. Some systems lack a few parts, depending on if it is a passive or active system, but all of the basic components are pertinent to understand.

Growing Chamber

First, you need a growing chamber of some sort. This is the part of the system where the plant roots will be growing. It is the container for the root system, if you will. This growing tray provides plant support and stabilization, and it also protects the root system that is growing from pests, damaging heat and sun rays, and too much light.

What is important about the root zone is keeping it cool and void of as much light as possible, and the growing tray can help with that. If exposed to the light for too long, or heated by the sun too much, the damage the root system will sustain will compromise its ability to absorb the nutrients from the water you are providing it, and this eventually leads to plant death.

The size of the growing chamber simply depends on the type of system you will be utilizing and the plants you will be growing. Bigger crops have bigger root systems and, therefore, require bigger growing trays than smaller plants with smaller root systems. After all, a basil plant and a watermelon will need different forms of support and yield different sized root systems. Just make sure that whatever you use for your growing tray will not react somehow with the nutrients or corrode in any way because of the water.

Reservoir

A reservoir is also necessary. This is what holds the nutrient solution you create with your water. Depending on the type of hydroponic system you wish to utilize, you may or may not need some sort of submerged pump that will shoot the nutrient solution to the growing chamber on some sort of a timed cycle.

But, the reservoir can also simply have the root systems dangling in it all the time. In this particular case, the reservoir also acts as the growing chamber.

Anything that holds water, and does not react negatively to the nutrients it will be filled with, can work as a reservoir. As long as it does not leak and holds the water amount you need, you will be good to go. Just make sure to clean it out well beforehand.

Keep in mind, however, that your reservoir needs to be light-proof, which is why plastic is the most common element used for hydroponic nutrient water reservoirs. But, if you do not want to make it with plastic, it is easy to make something light-proof simply by wrapping it in something like bubble wrap or painting over the outer surface altogether. The reason why it needs to be light-proof is because microorganisms and algae that will damage your nutrient water can begin to grow with even low lighting levels.

Submersible Pump

If your system is active in any way, then a submersible pump will be necessary as well. This is what will help get the nutrient water to the growing chamber your plants are nestled in. These types of pumps can easily be found at any hydroponic supply shop and home improvement stores (in these stores, they can be found in the garden supplies and pond pump section). A pump is nothing more than an impeller, which is a circular device that moves via fluid rotation, that utilizes an electromagnet to spin it. If your system is an active one, then this is the piece of the puzzle that makes the hydroponic system you have chosen "active."

PVC pipes and connectors

Your hydroponic system also needs a way to deliver the nutrient solution. Simply having the submersed pump does nothing, there has to be a connector of some sort that the water can flow through in order to be delivered from the reservoir tank of water up to the growing chamber in the first place. The most popular things that are used in homemade hydroponic systems are a series of PVC pipes and connectors. But, many people also utilize irrigation tubing and connectors as well as blue and black vinyl tubing. As long as it is connected tightly, does not leak, and delivers the water in the necessary way when the pump kicks on, then there is nothing else to worry about.

Sprayers and Misters

However, if you are using a system that recovers the used water and stores it back into the reservoir, make sure you have the same type of tubing that not only gets the water to the plants, but drains the water back off and into the original reservoir. Not only that, but some systems you can choose from use misters and sprayers instead of the "flooding" that the basic tubing provides. This means another piece will be involved from connecting the tubing to the pump and getting the water to the plant.

Many hydroponic stores will have a vast array of sprayers and misters, and the sales associates can talk you through the types of piping that go with those various misters. Be careful, however, because they can clog easily, especially if you are recovering the water and recycling it. Make sure to have extra ones handy for when clogs do happen so you can switch it out easily and in a timely fashion.

Timer

A timer is another requirement in an active hydroponic system. Some systems only require one timer, and some require two. One timer is for the submerged pump, and the other one would be for any artificial lighting that might be required. The timer for the lights controls the on and off functions, and the timer for the pump controls when the pump is cut on and off, which either pumps the nutrient water through the piping and onto the plants or stops the pumping and drains the excess water out of the growing chamber.

One thing to keep in mind with timers, however, is to get one that is rated for 15 amps instead of 10 amps. 15 amp timers are heavy duty timers, and they can withstand a great deal more than 10 amp timers. Also, they are usually water resistant, which will be imperative for the job it has to fulfill. However, I recommend sticking with an analog timer for a couple of different reasons:

1. they are cheaper than digital timers, and

2. they do not lose their internal memory settings if they lose power for any reason.

If you want to invest in digital timers, make sure you get one that has a backup internal battery in case you lose power to your hydroponic system. That way, you do not have to go back and reset all your timers again and again.

Air Pump

Another necessity to a hydroponic system is an air pump. Air pumps can be found at any aquarium supply store, and what it does is it simply supplies the root systems and the nutrient water with air and oxygen. How it works is this: air is pumped through a line to a layer of air stones, and these stones create many different small bubbles that rise up through the nutrient solution. However, in a water culture system, the air pump serves a different purpose: it keeps the plant's root system from suffocating while they are submerged in the nutrient water at all times.

No matter what, however, this moving air and oxygen not only oxygenates the water, but it keeps the nutrients moving around and circulating, which helps greatly in keeping everything evenly mixed. This moving water also helps reduce the chances of pathogens and algae from being able to gain a foothold in the reservoir and begin multiplying.

Grow Lights

Grow lights are optional and depend on what you are growing with your hydroponic system as well as where your hydroponic system will be set up. Natural sunlight will always be the preferred "grow light," but that is not always convenient or doable for some people and their environment. But, if you are growing specific plants during a time of the year where sunshine is not always available, then grow lights will be necessary in order to sustain healthy plant life.

Do not confuse a grow light with a standard house light. Grow lights have specific color spectrums they emit that mimic natural sunlight and moonlight. These wavelengths of color are then utilized by the plants to conduct their process of photosynthesis. This is the process they need to undergo in order to produce budding flowers and fruits. All this means is that the type of grow light as well as the amount of light a plant receives will affect the main process by which they grow. If you want to purchase grow lights for your hydroponic system, they can be found in many gardening areas in home improvement stores.

Water Culture System

A wonderful entry level hydroponic system is the water culture system. It is easy to maintain and gets you familiar with many of the working parts in a basic hydroponic system. To set up a water culture hydroponic system, you will need the following items: a reservoir, a slab of styrofoam, the plants you wish to grow, an air pump, an air stone, the proper tubing, net pots, and your desired growing medium.

Now, follow these easy steps:

- Find a reservoir to use for your water nutrient solution.
    - Make sure it has the same dimensions on top as it does the bottom (i.e.: if the top is 36 x 40 inches, then the bottom should also be 36 x 40 inches. No tapered reservoirs!)

- If the reservoir you have picked is translucent, clean it well, let it dry, then spray paint and leave out to set.

- After the paint has dried, use a measuring tape to obtain the length and width of the container.

- Once the dimensions have been taken, grab your slab of styrofoam and cut the dimensions ¼th of an inch under.

    - As in, if the dimensions are 36 x 40 inches, then cut the styrofoam at 35.75 x 39.75 inches.

    - Also make sure to carve out a little semicircle at one of the ends of the styrofoam so the air tubing can sit nicely to the side without becoming compromised.

- Then, take your net pots and place them on the styrofoam and draw lines around the bottom. These circular lines will designate where you cut the styrofoam out so you can place the net pots into the styrofoam. Do not trace them too big, otherwise the net pots will slip right through them when placed inside the holes.

- o Remember to space these net pots well so the root systems do not tangle and the plants do not suffocate each other as they grow.

- Now, it is time to begin hooking up your submerged pump. Connect the airline to the pump and attach the air stone to the free end. Put the air stone in the reservoir, fill it with the nutrient water mix, then turn it on.

  - o If working properly, you should see bubbles rise to the surface.

- Next, take something sharp and cut out the styrofoam holes you traced earlier. Place your net pots into the styrofoam holes, making sure the net pots will be secure in the holes.

- Place the entire styrofoam/net pot arrangement slowly down into the nutrient water, turn the air pump on, and you now have your first fully-functioning hydroponic system!

Building your own hydroponic system is best done by practicing on smaller systems like this. If you are handy and have had many years of experience putting things like this together, then you can start out with something bigger. However, for the beginner hydroponic grower, this system is easy to put together, easy to maintain, helps you perfect your craft, and is easy on the wallet.

However, putting it together does not mean it is finished, because a hydroponic system also requires maintenance.

Chapter 6: How To Take Care Of Your Hydroponic System

Key Takeaway: *Just as your plants require maintenance, so does your system. Keeping it sterile, clean, and constantly checking on its systems will ensure things are working properly and that your plants will thrive in the environment.*

With a hydroponic system, a certain amount of maintenance comes with it. Some systems will require maintenance that is not necessary for others, but all of them have a certain amount of upkeep. For example, a passive system does not need any worries over electrical outlets and moving parts, but the pH balance for the nutrient water is essential. These types of maintenance differences can mean the difference between your plants flourishing or dying, and it is important to know the specific maintenance requirements for the particular system you have chosen.

Likewise, if you feel that the maintenance that comes with the system you have picked is going to be too much for you as a beginner, then do not worry about it. Simply use the maintenance guidelines as a way to gauge what you can handle now, then use the system attached with those maintenance guidelines to determine which of the 8 easiest fruits and vegetables you will grow first.

How you get to the system of your choice does not matter, what matters is making the best choice for your lifestyle and skill level that will help you instead of hinder you.

Testing the Nutrient Balance

With every system, a certain amount of upkeep on the nutrient water will come with it. Testing the nutrient balance as well as the pH balance of the water should be done every three days. However, if you are utilizing a recovery system, it should be checked every single day. Paper test strips are the most inexpensive way to check the pH of the solution, but many people enjoy using a digital pen. It is more expensive, but it is much more accurate. In terms of checking the nutrient balance in the water solution, purchase a EC (electrical conductivity) meter in order to measure the strength of the nutrients in the water solution. If it is too strong, add a bit of water. If it is too weak, add a top-off nutrient solution. However, do not simply "top-off" more than three times before completely changing the water in the reservoir.

But, when it comes to completely changing the water in the reservoir, it needs to be changed every three weeks, no matter what.

And, as always, make sure to retest the pH after adding water or topping it off.

Check the Water Level

Make sure to also check the water level in your reservoir. This should be done every three days. Then, when you need to add water to your system, simply mix the amount of water you need with the amount of nutrient solution it requires (many nutrient additives are mixed together in a specific amount based on the water volume it is being mixed into), put it into the reservoir already mixed, and then measure the pH levels to make sure they are alright. Adjust accordingly, if necessary.

Check for Growing Patterns and Pest/Diseases

Another maintenance tip is to check your plants daily for growing patterns and check their buds and leaves to make sure no pests or diseases have attacked your precious plants. If you some anything wrong, do what is necessary to fix it immediately. Any threat that sticks around for any length of time heightens the risk of plant death. Write everything down in a growth book, as well as observations and any techniques that needed to be implemented that day, and date it. This will help you to be able to look back on things and find patterns with your growing potential, which can be useful if you want to perfect your craft or find patterns in how your plants grow in the environment you have them in.

Clip Plants

When plants are full-grown, instead of harvesting the entire plant you can simply clip the outer parts of it (like with herbs, lettuce, and spinach) instead of harvesting the entire thing. This not only keeps a constant supply coming, but it reinvigorates the plant. The only reason this would not be acceptable is if you allowed the plant to flourish for too long. When some plant life reaches a certain point, their own size can be their detriment. So, you would harvest the entire thing and leave the root system intact and ready to grow another plant life from its foundation.

How do you know if something needs to be wholly harvested? Some of it is personal preference, but many hydroponic farmers clip whole leafy vegetables from their gardens if they have to take more than 5 or 6 clippings from the plant in order to trim it down to size. For example, if you have a head of lettuce that has flourished and you want to trim it down, that is fine. But, if you have to take more than 5 or 6 full leaves off the plant to trim it down, many farmers simply harvest the entire head of lettuce and start fresh with that same root system foundation.

Be Aware of High Levels

Always be aware of light levels! Whether you are using actual sunlight or artificial sunlight, it can have detrimental effects if not given enough or given too much sunlight. Likewise, wind needs to be taken into consideration. If you have your plants near an open window and there is a strong breeze, that could damage your delicate root systems and growing plants. Judging these elements, you have your plants exposed to allows you to adjust outside items necessary to combat things, such as adding or removing screen cloth or even moving the entire system to a more protected area. But, the most important one is whether or not your plants have been in the rain. If so, check the reservoir immediately for nutrient density and pH levels. This dilution, if not fixed quickly, can lead to root system death.

Check Electricity When Using an Active System

If you are using an active system, then making sure the electricity is flowing as necessary is crucial. Always check to make sure the air pump and air stone is working as it should. Also consider having some sort of backup power specifically hooked up to your hydroponic system, in the event of power failure when you are not around. If the power fails, your plants cannot get the nutrients they need, which may result in plant death.

Pipe Maintenance

There is also maintenance that come with the piping that hooks up your entire system. Once every month or two, unhook the piping and flush a mixture of distilled water and peroxide through its tubing. This will keep the pipes clean and free of buildup that could contaminate the nutrient water and clog up the lines. And, speaking of cleanliness, whenever you completely change out the water in the system's reservoir water tank, make sure to thoroughly wash and rinse it down. If you have a misting system, soak those in a water and peroxide solution and flush water through them to make sure the misting spout is not clogged.

Overall cleanliness is essential to any system such as a hydroponic garden.

When you change the nutrient solution out completely every three weeks, there are some things to keep in mind that have to happen: first, you have to make sure the main source of electricity (if being utilized) is completely disconnected. Then, when removing the plants, make sure to wrap them in something damp, like a wet paper towel, so they do not dry out while you are cleaning and changing the reservoir tank. This should be done whether or not the plant has a growing medium, but it is required if the system you are using does not require a growing medium. If you want to completely recycle the water and have outdoor plants, consider dumping the nutrient-rich water on top of them to fuel their own growth patterns.

Use a Calendar

These maintenance concepts are crucial to the cleanliness of any hydroponic system. But, they are also crucial to the development of the plant life you are attempting to grow. If you need to, use a calendar specifically for your hydroponic garden and put it on the wall somewhere where you can see it every day. If you run your calendar electronically, input everything you need for the coming months into your device and add some ringers and bells to it. Reminding yourself of these tasks is essential to the development of not simply your skill, but your crop yield as well.

However, outside these basic maintenance requirements are some tips and tricks that can be utilized when growing any plant in any hydroponic system you choose. These are the types of things that will help tailor your plant life to its current surroundings, give you some things you might not have thought of yet to implement, as well as help make this learning process more cohesive and less overwhelming.

Chapter 7: Tips and Tricks For Success

Key Takeaway*: If you are feeling overwhelmed, don't be. Creating your own schedule to help remember things will aid you in the long run. Until then, there are many different tips and tricks formulated by other experienced farmers that will help you while you cultivate your own techniques!*

Even with the outlines and the instructions, hydroponics can be very overwhelming for beginners. As long as you take it step-by-step and give yourself time to learn the tricks of the trade, there is no reason why hydroponics couldn't be the new wave in how many people keep their own personal gardens.

But, if you are feeling overwhelmed, there are some tips and tricks that can help you get used to this new way of farming.

Dangling Roots

First, if you are utilizing a system that has dangling roots, like a passive system or the nutrient film technique hydroponic system, trimming the roots back every so often will help to keep pipes unclogged and roots from tangling. When roots tangle, they apply pressure, and this increases the chance of root system death and plant drowning. Keep in mind, however, trimming roots is not like trimming hair. When you do go to trim the root systems back, only $\frac{1}{8}$th of an inch or so from each root is necessary.

Research

Research, research, research. For a beginner, most of your time will be spent on research. Talk to people who have been doing hydroponic gardening for a while. Utilize social media groups for their information. Find a hydroponics store and plan a trip to go talk to some of the employees. The more research you can have under your belt before beginning, the less painful learning is going to be in the long run.

Nutrients Chart

Create a chart for the nutrients every plant needs. For many, they already know which plants they wish to grow and harvest, and some plants have different harvesting times than others. Make a homemade chart that lists the plant, the required nutrients, the time it takes them to grow hydroponically, and how they should be harvested. Writing down that information yourself helps commit it to memory, and then having it somewhere close helps establish a point of reference, should you need it later along the line.

Other things that should make it onto this chart are sunlight and darkness cycles, if necessary, as well as the plant feeding schedule. Knowing this information will help you avoid plant death during your first harvest.

However, if the nutrient percentages become too confusing, then take the easy route: use a professional three-part nutrient product. Hey, if it is good enough for NASA, it is good enough for us! But, stay away from the nutrient additives your first few harvests. Get a hang of basic nutrient solutions, and how to change them out, before adding more complicated matters to the mix. Trust me, it will help a great deal in the long run.

Even though this seems like common sense, make sure you have all of the necessary nutrients and equipment before you begin. Check, double-check, then triple-check that you have everything. The worst thing you can do is start putting your highly-vulnerable plants together only to realize you don't have all the components.

Minimize Artificial Light

Try to keep your hydroponic garden in an area of the house that has the least amount of artificial light. Minimizing fluorescent light exposure will help aid plant growth as well as help regulate their photosynthesis actions. Just make sure they are near a natural light source, such as a window, in exchange for getting rid of that fluorescent light source.

Another substance that needs minimal light exposure is your nutrient solution. Remember when we talked about painting your reservoir to make sure light cannot penetrate to the solution? Bacteria will grow in harvesting numbers if light touches that solution for long enough. Keep it out of all light sources, period.

Only Use Nutrients From Reputable Brands

Yet another thing to keep in mind is making sure you do not purchase off-brand nutrients. It is better to purchase your plant's nutrients from a reputable hydroponics gardening supply chain so that you have their neighborhood reputation to back up their product. Purchasing an off-brand product can lead to many different issues, from constantly imbalanced pH levels to the wrong percentage of base nutrients, and these issues can wreak havoc on your crops without you ever realizing what you are doing. Stick with the hydroponic retailers when it comes to your nutrient solution.

Sterilization

The rule of thumb for sterilization is to do it between crops. Unless your pipes or misters become clogged, sterilization and a complete clean-down of your hydroponic garden should happen in between major harvesting sessions.

When adding new plants to your garden, many people that utilize a root-dangling technique allow their plants to grow in soil for up to two weeks. Some even allow the plants to grow as long as a month. Whether you grow them yourself or purchased them already grown, make sure the root system is not so premature as to succumb to being drowned in the nutrient solution.

Be Careful With Pollen From Another Garden

Something that many people do not think of is how pollen from another garden can affect yours. If you have visited another garden and arrive home, change your clothes and wash down before tending to yours. That foreign flora pollen can alter the growth cycles of your plants as well as germinate them in ways you did not expect. Hydroponics gardening gives the farmer total control over the different elements, and pollen from another garden are no different. If you visit another garden, get your clothes in the washer and take a shower before tending to yours.

Visitors to your garden should also follow many of these same rules. You have worked hard for your hydroponic garden, protect it!

Invest in a Screen or Filter

If you are doing your hydroponic garden outdoors, consider investing in a screen or some sort of filter to put over your air intake and exhaust system. It will protect those inside your home as well as your garden from infiltrants that can wreak havoc on allergies as well as plant growth.

Keep a Grower's Journal

Something that might help you as a farmer is to keep a "grower's journal." This is essentially a diary that charts everything you see and do on a daily basis. Some people do it to simply keep track of what they have done to their plants, and some people chart everything from how their plant looks to what exact percentage of nutrients are in their water solutions at any given point. This can help you keep track of all the variables you wish that go into your hydroponic garden, and it can give you a stable and historical reference after you have been gardening for a while. Any change or observation you make should go down into this journal, and the more you date this journal the more you will be able to predict growing cycles, harvesting sessions, and when things like changes in nutrients is necessary. This journal can help you completely tailor your entire schedule to the needs of your garden, and it will help you pinpoint exact factors that are needed in helping your plants grow to their fullest capacity.

Check Water Levels

One of the biggest factors in hydroponics gardening is your water, which is why it deserves another look. If the pH balance is off, your plants will suffer. If there is too much water, your roots will rot. If there is not enough water, your plants will dehydrate. Out of all the parts in the equation of a hydroponics garden, the water is what you need to tend to the most as a beginner. Eventually, it will become second nature, but until then the water levels, both in amount and in nutrient, need to be gauged very closely in order to reap the benefits of that first satisfying harvest.

Pruning Shears

Also, make sure to understand the incredible capacity of pruning shears. If you see a leaf or a stem that is useless or rotting, prune it off before it begins to affect the entire plant. Things like this can rob the plant of valuable energy and nutrients which can affect the overall harvest. Likewise, if a plant is becoming too big, pruning shears and cutters can be utilized to cut off a few leaves and pluck a few herbs so the plant can keep growing healthily. The "cleaner" your plant can look and be, the bigger your yields will be without having to replant in the long run.

Nutrient Solution Temperature

Make sure your nutrient solution temperature is also acceptable as well. Overly hot or overly cold water can affect the nutrients that are intermingled with the water and, in return, those drastic temperatures can have damaging effects on the root systems they are trying to nourish. Keep this solution between 65 degrees and 75 degrees Fahrenheit to ensure optimum plant life and nutrient absorption.

Conclusion

Thank you for taking the time to read this book, '**Hydroponics 101**: *The Easy Beginner's Guide to Hydroponic Gardening. Learn How To Build a Backyard Hydroponics System for Homegrown Organic Fruit, Herbs and Vegetables.*'

If you have read through it and digested it, then you should have a clear understanding of the basics of hydroponics. I urge you to keep this as your own personal manual, not only to answer questions but to also guide you. Many people out there will try to complicate hydroponics and add in things that are not necessary, and when you have something that harbors a solid foundation, it is easy to come back to and rely on.

In this book, we covered all the different hydroponic systems, what makes them different, what makes them useful, as well as different beginning plants to start out with as well as every single basic component a hydroponic garden could have in case you feel the want to build your own. We scaled the world of maintenance and how to upkeep your plants and clean your hydroponic system, and we even gave you some tips and tricks to implement over the coming months as your skills begin to bud and you start to commit this information to memory.

The thing you need to understand is that this takes practice. At first, it feels overwhelming and intimidating, but do not allow it to swallow you hole. Use the outlines in this book and create your own checklist of things to do. Just like learning a new sport or acquiring a new skill, memorizing the rules comes much easier than many think. The how-to hydroponic garden laid out in this book is perfect for beginners to assemble in order to practice the basics and cultivate their hydroponic farming technique.

Feel free to branch out and do your own thing. These basic components can be strung into many different combinations, and many different combinations of plants can be grown in one hydroponic system. Make one for your classroom or make one that covers your entire backyard. The possibilities are endless! There are those who do windowsill hydroponic gardens to simply grow fresh herbs, and there are those who become so skilled in this art and technique that they stop relying on grocery stores for their produce altogether. The world is your oyster with this type of project, and the only limitations are the ones you impose upon yourself.

If you enjoyed this book, please take a second to leave a review of the book. I appreciate any honest feedback, and it helps in my ability to continue producing educational books that take broad subjects and breaks them down into bite-sized pieces.

Welcome to the world of hydroponics. Your gardening will never be the same...

BONUS CHAPTER: What is Aquaponics?

This is a bonus chapter from my book '*Aquaponics 101: The Easy Beginner's Guide to Aquaponic Gardening: How To Build Your Own Backyard Aquaponics System and Grow Organic Vegetables With Hydroponics And Fish.*'

Enjoy!

***

Key Takeaway: Aquaponics, the combination of aquaculture and hydroponics, is changing the face of farming and the growing of vegetables. It can be achieved by anyone willing to learn and has great benefits far outstripping those of regular gardening and hydroponics.

Defining Aquaponics

There aren't too many practices that, when combined, can expand on the advantages of each other while reducing and in some cases eliminating the limitations of each. Aquaponics can be thought of this way. For this reason, it will truly bring your garden to life in more ways than one. You may be thinking what exactly is Aquaponics, though it is really quite simple yet ingenious.

Aquaponics, to put it simply, is the coming together of aquaculture (the raising of fish) and hydroponics (growing plants without the need for soil). This ultimately creates one integrated system that provides food and nutrients for fish who in turn create waste that, through microbes and worms that thrive amongst the environment, convert that waste into fertilizer for the plants to allow them to continue providing the fish with an abundant environment.

This occurs as the microbes or nitrifying bacteria and red worms convert ammonia from the fish waste into nitrates, as well as the solids into vermicomposting, creating a viable food source for the plants.

There are three methods used to establish your aquaponics system: media based, raft based and a hybrid combination of these two.

1. **Media Based** – The name comes from the media (gravel, clay, pellets and other material) in which the plants are grown. The media allows for the ammonia based waste and mechanical waste to filter through. This means there is less work to do to maintain the system. The media based system is often more suited towards growing larger plants that provide fruit.
2. **Raft Based** – The raft based system makes use of a foam raft that floats in a channel accessible to the filtered fish effluent water. The plants are secured in holes along the raft, while the roots have access to the water below. The raft based method is suitable for growing smaller plants,

such as salad greens and other plants that require less nutrients to thrive.
3. **Hybrid Method** – Using a combination of the two, the media beds are able to filter the solid waste prior to the water entering the raft system. This is ideal as it provides further flexibility in the planting process and means much lower maintenance. While this system may be more expensive to establish, the benefits allow you to have a wider range of what you grow in your garden and a much higher yield.

Advantages Over Traditional Gardening

Soil based gardening not only takes additional effort and maintenance, the practice is rife with drawbacks that leave your garden a constant source of attention without providing a reliable return. Traditional gardening through the use of soil requires constant watering, particularly in hotter climates.

This will take effort on the part of the individual who has to water the garden, either yourself or a paid gardener, or an irrigation system, which can be quite costly to both establish and use. This doesn't take into account the fact that you will need to constantly monitor your garden for weeds which can crop up over time, draining precious resources and nutrients from the plants you had placed so much effort in providing an environment in which they are able to thrive.

Then comes the insects that live amongst the soil. Going out to admire your garden can come with a shock when you discover that your plants are under attack from a pest that is slowly destroying them. These pests are not just limited to insects, they can come in all shapes, sizes and species. Think rabbits, raccoons, or anything else that finds its way into your garden to feast on your hard work.
While some people enjoy the time they spend working their garden, it is not always best for your health. The more effort you extend towards your garden, the more strain you may be placing on your body such as your

back, knees, hands as well as the additional risk of being exposed to the sun. Over time, and as you maintain your garden, this damage can become significantly noticeable. However, this is all avoidable with aquaponics.

If you feel as though aquaponics may have some learning curves to overcome or that establishing your system may be more complex than you had originally thought, this is no truer than the case for traditional methods of soil gardening. With traditional gardening, you need to learn the correct timing for:

- when to water your garden
- how much is required
- the fertilizers to use, and
- what soil composition you need for which plants.

How Does Aquaponics Differ From Hydroponics?

You might be thinking that these issues can be overcome with traditional hydroponics, the growing of plants without the need for soil. Traditional hydroponic systems can be quite costly and require additional chemicals, sales and other trace elements to create the right environment for the plants you are growing.

These mixtures also need to be carefully analyzed to ensure that the correct pH level is adhered through the use of additional meters and tools. On top of this, the water in your hydroponic system require replacement frequently due to the buildup of sales and chemicals which become toxic to the plants that rely on them. This presents an entirely separate issue on its own, the disposal of this mixture needs to be environmental considerate.

Even more alarming, hydroponic systems are prone to a disease which can be devastating to your plant life. The disease is known as "Pythium", or root rot. While it can found in traditional hydroponic systems, it is extremely rare in aquaponics.

The case for aquaponics is that your fish can be fed anything from just standard inexpensive fish food all the way to food scraps from the house that you want to dispose of. This means you have a reliable system for composting that doesn't require additional harmful chemicals and salts.

The monitoring of your aquaponics system consists of nothing more than testing the pH and ammonia levels on a weekly basis and of course if there are any visible issues with the fish in the system or the plant life. It is advised that you keep an eye on your system during the initial month or so to ensure that operations are running smoothly. This also provides an opportunity to correct any issues before they destabilize the system down the track.

The water system in your aquaponics set up doesn't require any discharging or replacing. The only reason it would ever need to maintained is during hotter months or when the water evaporates naturally in which case you will only need to provide a top up to keep the fish and plant life content.

Why Aquaponics is Relevant Today

As our society moves towards more renewable and sustainable methods of growing food, the concept of Aquaponics was developed through research established in the 1970s spearheaded by Dr. James Rakocy at the University of the Virgin Islands. It was developed as a means of finding a way for plants to act as a natural filter. From here, the benefits of aquaponics became evident, as it further evolved in the 1980s and 1990s, to establish the systems as we understand them today.

Aquaponics solves many of the previous drawbacks of traditional gardening as well as hydroponics. The reduction in waste has been an advantageous feature for farmers looking for a way to conserve their water resources while also having an environment to breed fish. Aquaponics is noted to only use around 10 percent of the water that similar setups in soil based gardening use, and even less than that of hydroponics. This more efficient use of resources is also evident considering that the set up relies on a consistent abundance of water. This means you can never underwater or overwater the plants. The same can be said for the availability of nutrients, meaning your system no longer requires consistent fertilization.

The toxic by-product of the chemicals used in previous systems was detrimental to the environment and unsustainable in the long run, particularly for larger scale operations. Aquaponics focuses on natural resources and by-products of the fish to provide all the nutrients required to keep the plants health and stable in their growth.

The time and effort used to maintain an aquaponics system is a stark contrast with other more traditional methods. Rather than straining the body, using precious and limited resources, you are able to focus on tasks such as feeding the fish and tending to and harvesting your plants.

In recent years, aquaponics has become an increased focus in commercial farming with a number of associations and communities being established to represent aquaponics in the farming industry such as the Aquaponics Association. The mission behind the association was to promote aquaponics and to continue the progress of the technology to support the systems for further growth and development through annual conferences, attracting practitioners of aquaponics from around the world.

***

This is the end of this bonus chapter.

Want to continue reading?

Then get your copy of "Aquaponics 101"!

Did You Like This Book?

If you enjoyed this book, I would like to ask you for a favor. Would you be kind enough to share your thoughts and post a review of this book? Just a few sentences would already be really helpful.

Your voice is important for this book to reach as many people as possible.

The more reviews this book gets, the more parents will be able to find it and learn how they can grow their own organic vegetables with hydroponics.

Thank you again for reading this book and good luck with applying everything you have learned!

I'm rooting for you...

By The Same Author

Notes

www.ingramcontent.com/pod-product-compliance
Lightning Source LLC
Chambersburg PA
CBHW052201110526
44591CB00012B/2033